THE PATTERN

FTX: The Illusion of Decentralization and the System Beneath the Story

Jamil Hasan

First Edition · 2026
Published by Crypto Hipster Publications

ISBN: 978-1-972991-03-9

Cover design by Jamil Hasan
Interior design by Jamil Hasan

Printed in the United States of America

For Gabby—a builder and a leader who didn't just believe in the system but helped shape it.

The Pattern

Most people think collapse is sudden. It isn't.

It feels sudden because you only notice it at the end—when something that looked stable stops holding. But by that point, the system had already been under pressure for a long time. The signals were there. They just didn't look like warnings.

They looked like growth.

More users.
More capital.
More activity.
More attention.

From the outside, those are signs of success. From the inside, they are pressure building against a structure that may not be ready for it.

That pattern shows up everywhere.

It shows up in markets, where cycles are explained after the fact but rarely understood while they are happening. It shows up in systems where complexity grows faster than comprehension. And it shows up in people who operate inside those systems and respond to incentives, not intentions.

This book is not about a single event. It is about the structure beneath events.

It follows four perspectives:

How pressure builds inside systems;
How structure hides and reveals risk;
How systems respond when forced to adapt;
And how behavior ultimately defines what those systems become.

Through builders working inside crypto infrastructure, this book moves past headlines and narratives to examine what is happening beneath the surface.

Because systems don't fail for the reasons people think they do.

They fail because they change, and when that change becomes visible, the outcome is already set in motion.

If you step back far enough, the pattern becomes clear.

And once you see it, you stop looking at moments—and start seeing the system.

Preface

I'm Jamil Hasan. People call me the Crypto Hipster—a name I gave myself, and one I've chosen to own. It started as a joke. A meme. Then it became a signal. Not rebellion for its own sake, but a way of seeing things.

Crypto moves quickly—often faster than understanding can keep up. Sometimes, a topic I spent several podcast episodes fully understanding in the first week completely changed the next week.

There is a tremendous amount of noise in the market and not enough uncovering of the true signals. I aim to change that.

Discussion about the price of Bitcoin, meme-coin hype, 1000X speculation, and narrative repetition never truly interested me. I wanted to find out what genuinely matters.

So, I started the Crypto Hipster Podcast in February 2021. Not to chase headlines. To have actual conversations with people building this space.

Founders. Creators. Thinkers...people trying to reshape systems.

Crypto Hipster became more than a podcast. Nine seasons. Over 580 conversations. Hundreds of written works.

Then I hit an uncomfortable fact:

Curating discussions is not the same as creating ideas.

For years, I documented conversations. Published transcripts of all my podcasts. Built an archive. A colleague told me that if I wanted to be an influential leader, I would have to do more than simply host conversations. I would have to move the conversations forward.

That person was right.

This book is the fourth of sixty-seven Crypto Hipster Curtain Calls compilations spanning two hundred eighty-one interviewees. It is the first part of that shift.

Curtain Calls differ significantly from my first 399 books. It is not a collection of transcripts. It's an interpretation—a reflection, a synthesis of conversations, ideas, and lived experiences. It's where my voice meets the voices of the people I've learned from.

Crypto is not just technology. It's about power. About identity. About freedom. And above all, about people.

My path into this space came not from comfort. For the past five years, I have fought an aggressive desmoid tumor. What began as a way to keep my mind occupied grew into something larger. It became a way to process the world and my place in it.

This book blends two things:

- The conversations I've had with builders of the digital economy.
- And the lessons I have lived myself.

Conversations with others. And personal memoirs interwoven with them.

You'll see both here. Ideas about decentralization and finance. About infrastructure and governance.

You'll also see stories. Resilience, failure, faith, relationships, recovery.

This technology does not exist in isolation. It reflects who we are. Too often, crypto gets reduced to price charts. The authentic story runs deeper:

- Whether these systems actually change anything or just recreate the same structures with a new label.
- Whether innovation leads to freedom or just a different method of control.
- Whether we, as individuals, choose to take part on purpose or just go with the current.

This book is my attempt to slow that down. To take hundreds of conversations and extract what matters most.

To connect patterns.

To ask better questions.

To share what I've learned. Not as answers. As a framework for thinking.
This is not only my story. It's an invitation...an invitation to think differently. To question more. To build with intention.

I hope that if you see yourself in these pages, this book will serve as an inspiration for you to write your own story with more clarity, purpose, and conviction.

Table of Contents

Before I Saw the Pattern

Some moments feel enormous right away. Others only make sense later, once you can step back and see what was happening beneath the surface.

Mine felt disconnected—travel, job changes, personal crashes—like separate chapters rather than one story. Later, they aligned—not as random events, but as pieces of a structure forming long before I noticed it.

When I was young, I was drawn to movement—not to being busy, but to finding a sense of where I was and what I was doing. Even then, searching didn't feel like an end state. I felt that something else existed beyond the immediate.

Travel was one way I followed that itch. Angkor Wat, the Great Wall of China, and the trails of Tiger Leaping Gorge weren't just places to see—they were attempts to feel something truer than my routines. I picked up a phrase then: "Come back different." It was about travel changing you. I did not fully get it then, but it implied that change was possible. And I needed change.

My work life was the opposite. Structured, stable, defined. AIG taught me systems at scale—how information moves and how decisions get made in large organizations through process, accountability, and continuity. Any deviation had limits.

There I learned how things run when the rules are clear. But I always felt a disconnect between the orderly world and the idea that something else was out there. As the gap widened, it did not change right away. It ran alongside another pattern I did not see until later.

Years earlier, I did not think I had a problem. That is how problems begin. There was no single moment that screamed for attention. Life worked.

Apartment, routines. From the outside, it all looked fine. The system ran, so why question it?

I told myself that it was enough. That story kept the system going. It made it okay to ignore the rot. What I was maintaining was not a life. It was a structure without connection, without energy, and without genuine engagement.

My apartment looked stable. Full of furniture. My emptiness was in how it mattered. Nobody came over. Nobody wanted to be there. It was a container, not a home. Isolation can feel like control. As long as I believed what I had was enough, I did not have to face what was missing. That let the system limp along while everything slowly broke.

The moment that showed it did not feel dramatic, but it felt like exhaustion. I had pneumonia and was physically wiped out. I couldn't keep up with the basic routines. Lying on the floor of my apartment, too sick to move, listening to dark, depressive jazz, I had no distractions, no movement, and no way to pretend everything was fine. In that stillness, the narrative fell apart.

I remember a clear thought. I had been telling myself I had a life, but I did not. I was not lacking stuff, but lacking connection and direction. Everything around me was just enough to avoid collapse, not to build anything real. The realization hit hard—it wasn't working. It didn't need a tweak. It couldn't be improved gradually. There was a total mismatch between what I was living and what I needed. The system could not continue.

A week later, I got sober. That was the response. But the collapse had already happened. The years after were about rebuilding as a structure came back, but it was very different. Discipline replaced drift. The system worked again, not by avoiding things, but by aligning with something sustainable. A reset let me build something real. For a long while, that held until a new pressure showed up. I found Bitcoin after I was laid off from AIG. At first, I understood it as a public database. That made sense to me. It hooked onto what I already knew, but that view missed the deeper stuff, which took time to grow.

As I moved into the space, the ICO era of 2017 hit, and everything changed. What had been growing became nonstop activity. As demand shot up, there were opportunities everywhere. The pace sped up past anything I'd ever known or experienced before. Boundaries disappeared. People across time zones constantly reached out to me. Projects overlapped. Expectations ballooned. And it was impossible to step away.

That environment creates pressure, not as one big thing, but as a steady pile of small things. Every request, every opportunity, every message stacked up. The system stretched to hold more. My sleep fractured, my focus split, and saying no got harder as the system leaned into constant engagement.

From the outside, it looked like productivity, and maybe success. But the structure behind it was unstable.

I did not see it at first because the pattern just looked different. Instead of isolation, there was nonstop interaction. Instead of stillness, there was movement. But underneath it was the same thing: a system running in ways it could not sustain. The recognition came in a hospital. After my second heart attack and stent surgery, the expectation was simple: stop, recover, step away. I did not. I dialed into a John Maxwell coaching call from my hospital bed.

Looking back, I see that I just kept up the behavior that had defined the system. People reacted fast to that. Why are you here? Why aren't you focusing on recovery? Their questions were predictable. What surprised me was that I had to be asked. Inside the system, what I did felt normal. Only when others pointed it out did the mismatch show.

Then something smaller happened from that hospital bed, but it mattered. I bought Litecoin with the last ten dollars in my bank account. It wasn't a strategic investment. It was the recognition that the way I'd been living was not sustainable. Not just for health, but structurally, as the system could survive by taking on more, but only up to a point. The realization was not new. That purchase made it obvious. I had seen this before, not in crypto, but in myself.

Different conditions. Different environments. Different surface behavior—but the same pattern:
Pressure builds.
Systems adapt.
Structure gets strained.
And eventually, the gap between what seems to work and what actually works gets closed.

That recognition did not arrive as an abstract thought. It came from experience. Once it was clear, it was hard to separate what I was seeing in systems from what I had already lived through. That is where the shift starts: from story to system. What makes that recognition strange is that it does not feel like a discovery—it feels like a repeat. The specific details do not match, but the structure just feels familiar in a way you cannot ignore once you see it. The environments were different; the variables were different; the external signals were different; but the internal condition, the way the system ran under the surface, was the same.

In both cases, the system kept going past the point where it was sustainable. What made it dangerous was not pressure itself, not demand, and not just stress; it was the system's ability to keep working while those things were already happening. The function kept going, and that created an illusion of stability. Behavior kept running without interruption, and that made it easy to assume the situation was sustainable.

Sustainability isn't determined by whether a system can keep going in the moment. It is decided by whether it can keep going as conditions change. In earlier times, the conditions were inward: isolation, disconnection, the lack of anything meaningful beneath the surface. You could not see those things right away, but they set the limits of what the system could carry. When those limits were reached, collapse was not surprising, but it was inevitable.

Later, the conditions were both external and internal. The environment demanded more. As the pace picked up, constant engagement was expected. The system needed ongoing input to keep its outward face. The more it managed that, the more it reinforced the behavior that made it unsustainable.

That is why some kinds of success are hard to read. They send the right signals. Activity goes up, opportunities look bigger, and engagement is constant. From the outside, these look like growth, like momentum, like progress, and suggest the system can handle more. But those signals do not show the cost of making them. They do not show what is needed to keep that state going.

In both situations, the cost was absorbed instead of managed. There was no obvious line between what the system could carry and what it was being asked to do. Instead of limiting exposure, the response was to boost capacity by effort. And with each bit—more time, more energy, more engagement—the system justified what was happening in that moment.

That kind of compensation works short-term because it lets the system keep going. But it hides the real problem, which is that as long as a system can compensate, structural change gets postponed. The focus stays on keeping output up instead of checking sustainability. That creates a condition where the system becomes more dependent on the very behavior that is wearing it down.

The dependency is not always obvious and feels normal. Earlier, isolation became the norm; not connecting did not register as failure because it fit how things had been running. Later, constant engagement became the norm. Not resting did not feel like a problem because the environment expected nonstop activity.

Normalization lets patterns continue, turns warnings into ordinary signals, and lowers the urgency of change by framing the current state as acceptable, even when it is not sustainable. Therefore, recognition is often late, not because signals are missing, but because they are read within a framework that assumes stability. The system keeps working, and that is taken as proof that the structure is sound. Only when functionality is interrupted, when the system is forced into conditions it cannot compensate for, does the actual picture reveal itself.

The hospital was that interruption. There's a difference between knowing something in your head and experiencing it in a way that removes the option to look away. Lying in a hospital bed after a second heart attack is not abstract. It is a direct confrontation with the limits of the system when the consequences of sustained behavior are impossible to rationalize. And even then, the instinct was to keep going, to stay connected, to stay engaged, and to act like the system could absorb one more hit without consequence. The instinct came not from a conscious choice, but from the pattern itself, from the habit the system had developed in how it handled demand.

That made the realization different. Not that I realized something was wrong, it was that my way of responding to pressure had not changed, even though the situation had. The form had changed, but the structure had not. That is where the connection and the recognition of alignment become unavoidable. The same sequence happened twice, in places that looked unrelated but moved by the same dynamics. Pressure went up; adaptation happened; compensation let the system run beyond sustainable limits; and then, finally, the gap between what was happening and what could be sustained was closed by collapse.

That closure was not optional. It had to happen. The system as it was could not keep going forever under those conditions. What came next in both cases was not a return to how things were before; it was a rebuild that brought something that had not been there before: recognition.

Recognition that systems do not fail at random. Recognition that signs of failure appear long before we admit them. Recognition that being able to continue right now is not the same as being able to sustain. And most of all, recognition that the pattern is not limited to one area.

It is not only about personal behavior. It is not only about markets. It is structural. That realization does not instantly change what you do, but changes how you see things. It gives you a different way to read what's happening—one that looks beneath the signals systems send, and makes you aware of the conditions beneath those signals, of the dependencies that hold them up, and of the limits on how long they can last.

That awareness is not constant and does not remove pressure or the pull of opportunity. But it gives a reference point and a way to judge whether what is happening fits with what can be sustained. It creates a pause between stimulus and response, a moment where the pattern can be noticed before it repeats. That pause is not always enough, but it is the difference between running inside a system and understanding it.

And it is from that place, where experience meets recognition, that the rest of this analysis starts. Not from theory, but from having lived the structure before it had a name.

What changes when recognition takes hold is not the presence of pressure, but how you see it. Pressure does not go away; it keeps coming in different forms, shaped by fresh places, new opportunities, and new expectations. It shows up as growth, as momentum, and as the natural expansion of something gaining traction. And since it often looks positive, it is easy to accept it without questioning what it creates.

Recognition adds another layer, which is that it makes it possible to see pressure not just as an opportunity, but as a force that needs management, not just absorption. A force that, if ignored, will reshape the system in ways that are not obvious at first but are decisive in the end. This awareness does not remove the need to engage or the desire to build or to join in or to chase opportunity. What it does is change the relationship to those things by moving the focus from constant reaction to careful evaluation.

Instead of reacting to every signal, every demand, every chance, a new set of questions forms. What is this creating? What dependencies are being built? What assumptions are being made? What needs to be in place for this to keep going? And how stable are those needs?

These questions can be uncomfortable. They slow things down and add friction to processes that might otherwise run smoothly. They challenge the story of nonstop progress by asking whether that progress can be maintained. And in places that prize speed and quick response, this friction can feel like resistance.

But it is not resistance; It is alignment between what is being done and what can be sustained, between current demands and the system's constraints, and between behavior and structure. Without that alignment, the pattern repeats because people do not change enough.

Recognition alone does not change outcomes. It creates the chance for change. That chance must be taken in ways that alter the system's structure and not just its surface actions. This is where understanding and application part ways.

Understanding the pattern brings clarity, while turning that understanding into practice takes discipline. It takes the willingness to act against immediate incentives, to step back when the system pushes for more engagement, and to prefer sustainability over short-term output. These moves are not always visible and not always rewarded the way constant activity is, but they are needed if the pattern is to be broken instead of repeated.

Even then the pattern does not vanish. It stays as potential. The forces that shape it—pressure, adaptation, dependency, behavior—do not disappear. They keep working, bending the system in ways that are subtle and big. What changes is the ability to see them, to notice when they are building, and to respond in ways guided by that seeing. That does not guarantee a different result, but it makes one possible.

In both cases, the collapse that led to sobriety and the collapse that followed overextension, the outcome was not decided by one choice or one moment. It was the product of many conditions piling up over time, shaped by behavior, strengthened by the environment, and limited by assumptions that were not checked until they could not be sustained.

Those conditions are not unique. They show up across systems and appear whenever growth brings pressure, when adaptation brings complexity, and when incentives push behavior without fully accounting for sustainability. They are not always bad, though they create a gap between what a system looks like and what it actually is. That gap makes collapse necessary.

Structure needs correction. When perception and reality drift too far apart, the system must reconcile them. That reconciliation can take different shapes, but it always moves from assumption to check, from belief to limit, and from continuation to adjustment. In personal terms, that shift feels like clarity. In system terms, it looks like a collapse. The difference is perspective, but the process underneath is the same.

Structure ties these experiences together; not the details, not the settings, not the results, but the structure that made them happen. A structure working under the surface, shaping behavior, nudging decisions, and setting the limits of what can be sustained. Once you see that structure, it becomes hard to ignore. It is not obvious all the time and shows up in ways you can spot eventually. It appears in pressure building, systems adapting to that pressure, dependencies that were hidden at first, and the final solving of those dependencies when they can no longer be kept.

This recognition does not mean you fully understand every system, but it gives you a starting point. It is a way of looking that moves beyond single events and toward the conditions that produce them. And it becomes a way to read what is happening that rests on structure rather than on story.

From that starting point, everything else follows. What begins as something personal does not stay personal. It spreads outward and applies to the systems we make, the places we work in, and the patterns that come from them. It offers a lens to examine those systems, not as isolated things, but as part of how complex structures change under pressure. This is where the line between story and system fades.

The story stays. It gives context, experience, and a way in. But under it, there is something steadier. A structure that shows up across forms, places, and results. A structure that does not need any story to exist.

And it is that structure—not the story—that determines what happens next.

What I was seeing wasn't isolated moments. They were signals of something deeper—patterns that only become visible when you step back far enough to see the system beneath them.

The System Beneath the Story

Collapse is rarely understood as it happens. You feel it, react to it, and explain it—but you don't fully understand it. Visible failure draws attention, but when something that looked stable stops holding, that moment is only the surface—the deeper process was already underway.

By the time collapse appears, the system has been under strain. The conditions that made it possible formed earlier, quietly, slowly, and without being noticed. The same is true with people: breakdowns rarely start at the crisis. They begin long before, in habits and assumptions nobody questions. Failure is less an incident and more a resolution—for both systems and people. It is where structure and reality are forced into alignment. What looks sudden is often the compression of a long build-up, while the hard part is not seeing the collapse but understanding it.

Systems don't fail at random. They fail in ways that mirror how they were built, how they changed, and how people use them. In digital assets, this matters a lot. The industry is framed by moments—cycles, crashes, breakthroughs, and the stories that define them. Each cycle sounds unique, and each crash sounds like a one-off. But underneath, patterns repeat.

During the last cycle, that pattern was harder to see because everything looked like it was working. Capital moved freely. Liquidity was everywhere. Platforms presented themselves as stable, credible, and institutional in tone. For many participants, the system no longer felt experimental. It felt established.

Interfaces were smooth. Yields were predictable. Risk was abstracted away from the user experience. Participation expanded beyond individuals to include funds, firms, and institutions, bringing the appearance of structure and legitimacy.

From the outside, it looked like maturity. From the inside, it was dependency.

The system was not being tested to its limits—it was supported by conditions that allowed its assumptions to hold. Liquidity masked fragility. Trust was extended without verification. Control concentrated in places that did not present themselves as points of failure.

When those conditions changed, the system did not gradually weaken. It exposed itself. The collapse associated with FTX was not the beginning of failure. It was when the system could no longer operate under its own assumptions.

What failed was not just a company, but the belief that the system behaved as it was described. The distinction between custody and ownership, between decentralization and control, stopped being theoretical.

It became visible.

What appeared to break suddenly had been under pressure long before. Systems scale, absorb pressure, adapt, and then they hit limits that weren't obvious at the start. Those limits are not always technical but structural. Structure shapes how systems behave under stress—who holds control, how risk moves, and what assumptions must hold for the system to function.

Often you can't see structure when things run smoothly because behavior looks steady, relationships seem solid, and outcomes make people think the system is fine. But stability like that is conditional on assumptions that aren't obvious.

When those assumptions are tested, the system doesn't always snap right away. It adapts, and adaptation looks like strength because it lets things keep working under pressure. Adaptation absorbs shocks and preserves functionality as conditions change. But this adaptation is not neutral. It brings trade-offs, shifts risks, and creates new dependencies. Those adaptations pile up, and the system changes with them.

The gap between what a system was designed for and what it becomes is not immediately visible. It grows slowly. Behavior confirms it, and the interpretation gets messy because the system still works. Transactions keep happening, markets keep moving, and people keep participating. From the outside, those signals say everything's fine. From the inside, they might mean something else.

That gap—surface behavior versus underlying structure—is where systems drift away from their design, which is where failure starts. Understanding that drift requires more than watching isolated events. You need a framework. The conversations ahead are not a timeline of one collapse or separate interviews; they are perspectives. Each looks at the system from a different angle. Each shows a different way it behaves under pressure, how it adapts, and how it ends up acting because of the people in it. Together they make a sequence of dynamics.

The first dynamic is pressure. It starts the change process and comes from growth, from more participation, from rising expectations, and from the need to scale beyond early guesses. Pressure is not always bad; in contrast, it often means the system works and draws attention and capital. But pressure also constrains and tests what the system can hold.

As scale increases, new tradeoffs appear. Throughput must rise, costs must be managed, and user experience has to improve. These demands aren't optional if the system is to grow. So the system adapts.

Josh Bowen works in this layer. He focuses on the infrastructure that helps systems scale under pressure: the parts that process transactions, store data, and coordinate interaction across networks. From his point of view, pressure is measurable. It shows up as throughput limits, latency, and the capacity to handle more demand. Systems answer pressure in different ways, with some just adding capacity, making components bigger or faster. Others split things up, making modular parts that run on their own while still contributing to the whole. Each choice introduces trade-offs—speed, cost, and decentralization—balanced differently depending on priorities.

While the basic dynamic stays the same, pressure doesn't vanish. It gets absorbed instead. And as it is absorbed, the system changes. New layers appear to boost performance, intermediaries are added for a smoother user experience, and mechanisms pop up to manage cost and efficiency. Each fix tackles a constraint and creates a new dependency. Over time, the system grows more complex—not just technically, but in how it can be understood. Complexity doesn't remove risk, but moves it around. That's where the second dynamic shows up.

When systems run under steady pressure and growing complexity, the gap between what people see and what's happening widens. The system keeps working, but the assumptions become less visible. During the previous cycle, that pressure was amplified by rapid inflows of capital and expectations of continuous growth, which pushed systems to scale faster than their original assumptions allowed.

People interact based on what they observe. Transactions finish, liquidity looks available, and interfaces respond. But those signals don't always tell you how the underlying structure is arranged. This is the setting where structural failure grows.

Ishan Bhaidani's perspective looks at this layer. He argues that the issue isn't a lack of information, but that the information is fragmented. Signals exist, but they're scattered, with financial bits in one place, operational components somewhere else, and behavior in another. Each signal on its own can be rationalized, and any oddity explained away, but together they make patterns that hurt you if you see them.

Recognition is the problem. The system keeps functioning. And while it functions, belief stays strong. This belief makes perception lag behind reality. People treat the system as sound not because it survived stress, but because it hasn't been forced to run without its assumptions. When those assumptions break—when belief gives way to verification—the system's structure is exposed. That exposure is what people call collapse, but collapse is not the start of failure. It's when failure becomes visible. The system hits its limit, and what follows is not a reset, but a response.

The appearance of stability during that period made fragmentation harder to recognize, as strong surface signals masked deeper inconsistencies.

The third dynamic is reconstruction. After failure, systems are rebuilt. The first instinct is to stabilize the system and fix what went wrong so the same collapse doesn't happen again. Oversight often increases, governance becomes formalized, and integration with existing frameworks speeds up. The aim is usually to make things more resilient, easier to read, and better matched to the environment.

Annelise Osborne looks at this phase. She studies how systems change after failure, especially as they tie into traditional finance and new technology. Tokenization, institutional adoption, and regulatory alignment are pitched as fixes and as ways to improve efficiency, broaden access, and lower risk. Those developments are real. They change how systems operate. But they don't always change the core structure.

Often, these developments turn old systems into fresh forms of the same system. Assets get tokenized, yet ownership rules stay the same. Access grows, but control often does not. The system becomes more efficient and more compatible with existing finance, but by fitting in, it also picks up the old system's constraints.

Continuity shows up here. The rebuilt system is not exactly the one that failed, but it's not brand new either. It is shaped by the same pressures and steered by the same limits. If those limits remain, the system develops within them. And that brings us to the final dynamic: behavior.

Much of what is being rebuilt now reflects responses to that collapse, with efforts focused on making systems more legible to institutions and more aligned with existing financial frameworks.

If pressure pushes systems, if structure sets the bounds, and if reconstruction restores them within limits, then behavior tells you what the system really is. Behavior doesn't follow stories; it follows incentives.

Raj Parekh focuses on this level. He looks at how people use the system, how capital moves, where attention goes, and what opportunities get chased during booms and stress. Things often dismissed as noise—memecoins, speculative frenzies, rapid shifts in attention—become signals. They show what the system rewards. When it's cheap to join and quick to make returns, people act to fit that. They move fast, allocating capital for opportunity rather than belief. They adapt in real time, but that behavior is not irrational. It's consistent and mirrors the incentives built into the system.

Under stress, these tendencies grow stronger. Attention narrows as capital flows to the same spots, and the system gets pushed to its limits.

Infrastructure gets tested, constraints reveal themselves, and design and behavior display their relationship.

This is where the system shows its true face, neither as described nor as intended, but as it is used. The behavior seen during the last cycle—rapid capital rotation, speculative focus, and attention-driven allocation—was not an anomaly, but a reflection of the incentives the system made possible under those conditions.

Taken together, these four views form a framework for understanding the system. Pressure starts the change process; structure sets the limits; response rebuilds inside those limits; and behavior shows the result.

The order isn't tidy. It's a cycle, where each phase feeds the next, and each pass carries parts of what came before. The system doesn't reset after failure;

it evolves. And through that evolution, patterns keep coming back.

This book isn't about one event or a pile of isolated developments. It's about the system under those events, the forces that shape it, the structures that define it, and the behaviors that reveal it. Understanding these dynamics won't make uncertainty go away. But it changes how you read uncertainty by shifting attention from single moments to repeating patterns and from isolated outcomes to the system that produced them.

This understanding clarifies the whole—not as a string of disconnected incidents, but as a coherent system, always adapting, always developing, and always revealing itself through pressure, structure, response, and behavior. Once you see that pattern, it's hard to look at the system any other way.

To understand what's happening, stop looking at moments and start looking at the system that connects them—through pressure, structure, response, and behavior.

Pressure Without Release

Systems do not fail alone. They fail under pressure that builds slowly—often invisibly—long before anything breaks. This is the environment Josh works inside—not a theoretical system, but one shaped by real usage, real constraints, and tradeoffs that cannot be ignored. "Pressure shows up as throughput limits, latency, and the system's ability to handle demand." — Josh

At first, that pressure does not appear negative. It presents as growth, demand, and progress. More users, more transactions, more capital, more attention. From the outside, these look like success, but from the inside, they are forces acting on a structure that might not be ready for them.

As that pressure builds, traditional systems respond by centralizing. Decisions speed up, control concentrates, and tradeoffs are enforced from the top. That lets the system adapt quickly, but it reduces resilience. When those systems break, they break hard because the control points are also the fragile points.

In contrast, crypto promised a different path. The idea was to design systems that distribute trust, avoid central control, and better resist failure. In theory, pressure could spread across participants instead of being focused on a few key spots. In practice, that promise becomes messier. Decentralization alters structure, but does not remove pressure. Performance, cost, and usability still apply, and users continue to behave in ways that reintroduce centralization.

Pressure does not vanish. It finds other routes. That's where builders working on infrastructure operate. They are not in some perfect lab, but work in environments where tradeoffs are unavoidable and every choice has consequences. The system must work under real load, not just in theory.

"Performance is not just about speed. It's about how the system holds under real conditions." — Josh

Josh Bowen works in that space, at the layer where pressure is not abstract but immediate. His work sits at a layer most users never see—where performance limits, system design, and real-world demand meet, but that shapes everything above it. It deals with how data moves, how systems talk to each other, how transactions get ordered, and how parts of a network behave under load. Performance limits there are not abstract; they are immediate and measurable. And the pressure there is constant.

There are two primary ways to scale under pressure. One is to add capacity directly. Make blocks bigger, push more throughput, build stronger infrastructure, and accept that the system will use more resources and design around that. This favors performance. It assumes users want speed and that the system must deliver it, even if that requires more centralized or heavy components. "Systems respond to pressure by making trade-offs—speed, cost, and decentralization can't all be optimized at once." — Josh

The other is to rethink how capacity is spread. Instead of making each part bigger, make the system modular. Break it into pieces and let different parts do different jobs. Store data differently, process it in stages, and design it so no single piece carries the whole load.

Because of this, these are not only technical choices. They are philosophical. One accepts concentration for performance. The other tries to keep things distributed, even when it gets complex. Both answer the same pressure and bring their own fragility.

At the core, crypto engineering comes down to one simple question. What does the system actually guarantee? People assume a blockchain guarantees data exists forever, that it is stored and always accessible. That assumption is not quite right. The real guarantee is narrower and says that at a point in time a group of participants agreed that certain data was present. That it was seen and recorded, that consensus enforced that agreement, and it was put in a block. It becomes part of the chain's history. Beyond that, guarantees weaken.

Storage is not endless, and access is not always simple. Data being in the system does not mean it is easy to retrieve or interpret or useful without extra infrastructure. This matters because it reveals a gap between how people think the system works and how it really works. Users often treat systems as if they are whole and absolute, but builders know they are not. "What looks simple to users is usually built on layers that are anything but simple." — Josh What looks simple on the surface usually depends on hidden layers beneath—layers that must run correctly for the system to do what people expect.

Those hidden layers are exactly where pressure builds. Sequencing—the ordering of transactions—is fundamental to any blockchain. Early designs tied sequencing to the base layer. Transactions were submitted, ordered, and completed in the same system. The base layer's speed set the network's speed, while its cost set the network's costs. As demand rose, that model showed limits as users wanted faster confirmations, lower fees, and a better experience. Waiting for base layer finality was too slow, and paying the base layer expenses was too costly. The system could not meet those needs without being changed, so the system changed.

Sequencers appeared as an in-between layer. They sit outside the base system, take transactions, order them, and make blocks faster. They give users near-instant feedback—a sign that the transaction was received and will be included. And they batch transactions to cut the cost of putting them on the base layer. From Josh's perspective, this is where design decisions stop being theoretical and start shaping how the system actually behaves under load.

But sequencers also introduce something else: a point of trust. Users no longer interact directly with a fully decentralized system, but with an entity that can order, prioritize, delay, or even censor transactions. The system's rules may limit these powers, but the intermediary still shifts the trust model.

This is not a flaw—it is a direct response to pressure. The system could not meet real-world needs without that layer. Speed and efficiency demanded it. The trade-off was made because otherwise people would not use the system. Once added, though, the layer becomes part of the structure as it shapes behavior. Activity concentrates around it, while other pieces organize around that point. It becomes a kind of centralization inside a system built to avoid centralization.

This pattern repeats as the system evolves. Every push for performance adds dependencies. "Every improvement in one part of the system usually creates pressure somewhere else." — Josh
Every effort to cut costs adds new abstractions. And every layer you add to fix one problem creates conditions for another. Complexity grows—not just in code, but in how the system is understood. "The more you scale a system, the more the complexity moves to places users can't see." — Josh
And transparency, often said to be a blockchain strength, gets harder to read.

The data is there, but the links between components are not obvious. A user's action passes through many layers, each with its own rules and failure modes. For most users, this complexity is invisible, but for builders, it is unavoidable. And inside that complexity, fragility appears, not because the design was bad, but because the system is evolving under constraints and must adapt constantly.

That complexity doesn't exist in isolation—outside forces add more pressure. Crypto does not exist in isolation. It is shaped by the wider economy—shifts in liquidity, regulatory moves, and the behavior of big institutions working at a different scale. The reality is that it is tied to the same forces that shape traditional markets. When those forces shift, capital flows change, risk appetite moves, and narratives shift. Projects once seen as viable are reexamined.

Systems built for one environment must work in another. That is a distinct pressure—economic rather than technical. And it raises a hard question that can't be ignored. Who benefits from the systems being built?

Within that environment, institutions bring capital, scale, and different incentives—returns, market share, and integration into existing systems. That is not automatically bad. It can mean wider adoption, better markets, and more legitimacy, but it brings competition, not just for users, but for control.

Already visible, perhaps crypto-native infrastructure will become the base for systems run by institutions that were not part of its early phase. Value built in one layer can be captured in another, and the system can evolve to reflect the priorities of those with the most resources. That pressure forces different responses, driving different reactions. Some builders refine the current system and squeeze more efficiency out of it. Others try to redesign it from the start, to remove central points and spread responsibility better.

Both have value and are needed, but neither removes the deeper dynamics. The core problem remains the same. How do you build a system that handles real-world demand without bringing back the structures it was meant to avoid?

There is no single answer. Instead, there are tradeoffs, one after another, each tied to a specific constraint and each changing the system in ways you might not see at first. Those tradeoffs add up. They shape how the system behaves. And under enough pressure, they decide what happens next.

As those tradeoffs accumulate, today's systems become more advanced, faster, more flexible, and more capable than older ones. But they also run under greater pressure, have more users, require more capital, and are burdened by more expectations and scrutiny. Although they will adapt, the question is what those adaptations ask for and what they change. Adaptation is not neutral. It moves the system, sometimes toward the original idea and sometimes away from it, often into something no one fully expected.

This is consistent with how Josh describes these systems evolving. Not through a single breaking point, but through accumulated decisions made under pressure.

And under that pressure, systems do not break all at once. It reshapes them, forces choices, creates compromises, and builds dependencies. Over time, it shows something important. Not whether a system can survive pressure, but what it becomes while surviving.

Systems do not fail because they are weak. They fail because they adapt. In adapting, they become something else. Pressure is hard to spot while it is happening because it rarely looks like a limit. It looks like an opportunity instead. More users means adoption; more transactions means utility; and more capital feels like validation.

Each signal confirms that the system is working. But systems are not tested by how they work in ideal moments; they are tested when those signals intensify, when growth outpaces what the design expected, and when usage changes and assumptions diverge from reality. That is when pressure stops being abstract and when tradeoffs stack up.

Design alone doesn't solve this tension; outcomes do. Deciding which systems draw users, which keep liquidity, and which survive stress are important factors in determining what stays. Not noble intentions. What comes out of that process is a hybrid: part decentralized, part centralized; part ideological, part pragmatic.

Under pressure, systems do not break immediately. They adapt, layer by layer, decision by decision, until the structure that remains is no longer what was originally designed. At that point, what appears stable is only the result of accumulated tradeoffs. And when those tradeoffs are tested, the system does not fail because it was weak. It fails because it has become something else.

Pressure doesn't just test a system. It reshapes it—and what that pressure reveals is not just how the system holds, but how it is structured to begin with.

When Structure Breaks

Systems rarely look fragile from the outside. They often appear strongest when they are most vulnerable. Growth, visibility, and institutional backing do not prove resilience. "Systems rarely look fragile from the outside. They often look strongest right when they are most vulnerable." — Ishan
And they often mean pressure is building in a place that was never meant to hold it.

This is the lens Ishan brought into the conversation. Not a focus on individual signals, but on how those signals connect—and more often, how they cannot connect in time. His perspective wasn't about spotting the obvious. It was about understanding why obvious signals are so often missed.

The hard part about seeing structural failure isn't that there are no warnings. It's that there are too many signals that make people feel confident while quietly pointing the other way. By the time FTX collapsed, it was more than an exchange. It was woven into the ecosystem's infrastructure. It became liquidity, access, credibility—and, most importantly, a story that crypto had become acceptable to mainstream finance.

That story changed how people looked at it. No longer just another risky player, it felt like a foundation. Once something is treated as foundational, people stop poking at it. Doubt gets expensive. Skepticism makes you feel alone, so belief is the easiest move. Information didn't disappear within that environment; data kept flowing, metrics stayed visible, and behavior could still be watched by anyone who wanted to look.

Information alone does not create understanding. It requires interpretation—and interpretation requires context. "It's not that signals are missing. It's that there are too many signals pointing in different directions." — Ishan

Around FTX, the story was growth, markets maturing, and risk becoming professional. That narrative did more than describe it. It shaped how people made sense of things. The narrative didn't remove risk; it reframed it. "The story didn't remove the risk. It reframed it." — Ishan

Some risks became easy to shrug off. Others became harder to spot. When executives left, people called it a strategic move, not a warning sign. When volumes shifted, it looked like market evolution, not decay. When spending shot up, it was aggressive growth, not something unsustainable. Each signal fit neatly into the broader story, with its own tidy explanation. The system never had to confront the pattern those signals formed together.

This is where Ishan's framing becomes critical. The issue isn't access to data. It's the ability to assemble it into something coherent before the system forces that assembly on its own terms.

What was missing wasn't information; it was synthesis. This made the FTX environment so difficult to read. Not a lack of information, but the inability to connect it in time.

You could see each signal and explain each one. Liquidity flows were visible, balance sheet moves were visible, and executives behaved in ways you could watch. Market activity was there. None of these things, taken alone, were enough to raise an alarm.

The system did not hide its behavior, but it spread it around. That spreading creates a mental barrier. Seeing a system means linking signals across layers. You must connect financial moves to operational structure, and outcomes to the dependencies that make them possible. But systems don't present themselves that way. "The system doesn't give you clarity. It gives you fragments." — Ishan

They hand you fragments, where each fragment is readable and makes sense. But how they fit together is not obvious.

Under pressure, the fragments get worse. Things accelerate—decisions, capital, and reactions all speed up, and people get rewarded for reacting to what they can see, not for rebuilding the map behind those signals. Synthesis becomes harder, and acting on partial information gets cheaper. And the system stops being critically evaluated.

People interpret the system through little visible pieces that reinforce the dominant story. Little things that should be tied together look separate. Things that should be questioned get normalized. This is how structural risk piles up in plain sight.

Systems can keep running even as the things that let them run grow narrower. Functioning is not proof of structural soundness. Transactions can settle, liquidity can look stable, and interfaces can respond, while the dependencies that make those outcomes happen get more conditional and more fragile. That creates a gap between what the system does and what it can handle.

The actions stay the same, but what allows those actions to keep happening becomes more constrained. That gap doesn't stop activity; rather, it fuels it. Every successful cycle looks like the system can take more pressure, even as the range of conditions it can survive keeps shrinking. Over time, the system becomes dependent on assumptions that are no longer tested.

Liquidity will be there. Counterparties will behave predictably, and collateral will keep its value. As long as the system keeps working, those things don't need to be rechecked. They get built into how people read signals. That's where structural failure starts. Not as a sudden snap, but as a slow pileup of conditional dependencies.

The system keeps working, but its margin for error shrinks. When pressure goes beyond those conditions, the system doesn't ease into a new state. It is forced to reconcile all its assumptions at once.

What follows is not just stress, but misalignment. The system keeps acting as if its old conditions are still true, even when they're not.

Ishan describes this as the point where systems stop being interpreted correctly but continue to behave as if they are. The gap between perception and structure is no longer small enough to ignore.

Activity reflects expectations formed earlier, while the structure that should support that activity has already shifted. Transactions happen, positions stay open, and capital moves, but the link between those actions and the system's real capacity gets shakier every day.

This is hard to spot because behavior lags behind structure. People act on what worked before, not on what has changed. The system looks consistent because its outputs stay familiar, even as the constraints shaping those outputs change underneath. Over time, you get a split. What the system does no longer matches what it can sustain.

The longer this split goes on, the more the system depends on conditions that aren't guaranteed. Stability becomes conditional rather than structural. When that condition finally breaks, the system doesn't slowly move to a new normal. It must face the gap all at once.

Behavior that once seemed fine becomes unsustainable, not because people changed what they did, but because the structure that supported those actions isn't the same anymore. Systems that let behavior drift away from structure will eventually have to reconcile that gap. When they do, stories won't decide the outcome. The system's limits will.

That doesn't mean collapse can be eliminated. But it can be understood. Understanding won't prevent every failure, but it changes how we see it, as it shifts attention from the moment of collapse to the conditions that made it possible. It reveals collapse as the visible end of a process that played out. And once you see that process, it's hard to look away.

The same conditions are still around, the same pressures still work, and the same patterns keep forming quietly, under the surface. With FTX, liquidity looked sufficient, execution looked reliable, and growth looked strong. Those observations weren't wrong, but they were incomplete.

Completeness means putting the pieces together. It means linking financial signs to behavioral ones, operations to structural dependencies. It means asking not just whether something works, but what must stay true for it to keep working. That kind of integration is difficult, not because the data isn't there, but because the system doesn't show it that way.

The system presents fragments—each readable and coherent on its own. How they fit together is not obvious. At first, doubt gets absorbed and confidence slips slowly. But erosion has a tipping point, which is when the system flips from being driven by belief to being driven by constraints.

In a belief-driven state, assumptions persist. In a constraint-driven state, assumptions get tested. That's when structure shows itself. And showing structure breaks part of the story because structure sets limits. Once those limits are hit, the system must contract or fail.

In tightly linked systems, contraction is hard as everything is coupled. Fixing one thing pushes problems elsewhere. Therefore, unwinding happens fast: everyone moves at once, liquidity gets pulled, exposures are cut, and risk is repriced. Individually rational moves taken together destabilize. A system that counted on coordinated behavior under belief now faces uncoordinated actions under doubt.

Coordination unravels, as does the system's ability to work. What follows is not a negotiation, but a resolution. Once a system moves from belief to constraint, it no longer runs on what people think is true. It runs on what can be verified. Every assumption that once allowed the system to work without friction becomes a stress point.

Liquidity must exist, not be inferred. Collateral must hold value, not just be accepted. Relationships must survive pressure, not just look stable. This transition compresses time. What was deferred gets forced into the present.

Dependencies that were manageable one by one are tested all together. The system cannot adapt slowly. It must reconcile its entire structure at once. That's why collapse feels abrupt. This is the core of Ishan's argument. Collapse is not an isolated event. "Collapse isn't the start of failure. It's when the system can no longer operate under its assumptions." — Ishan
It is the visible correction of a misread system.

That abruptness is not a property of the collapse itself, but a property of perception. The system did not become fragile at that moment; it was recognized as fragile while the conditions that made it fragile were already there. They had been hidden by activity that still looked consistent.

When the lens changes, the same signals look different. Strength becomes dependency; liquidity becomes exposure; and coordination becomes coupling. The system does not change, just the interpretation does. So, collapse feels sudden, but it is not. It is also not the start of failure, though it is the end of misinterpretation.

Once misinterpretation is gone, the system is judged not by what it seems to do, but by what it can sustain. There is no room for the story to steady itself anymore. Only structure remains—and structure does not negotiate.

When that structure no longer holds under its own assumptions, what follows isn't just failure—it's the system's attempt to respond.

Rebuilding What Failed

If collapse reveals a system's structure, response is how that structure is made acceptable again. It does not erase what was exposed or start from scratch. It reorganizes what is already there. "Rebuilding doesn't remove the structure. Rebuilding reorganizes it." — Annelise
It imposes order where instability appeared and attempts to restore confidence in a system that revealed its limits.

This is the way Annelise framed the rebuild. Not as a reset, but as a reorganization shaped by the same forces that defined the system before failure. Her focus wasn't on what was added after the collapse, but on what remained unchanged throughout the response.

The instinct is not to ditch the system. It's to stabilize it, turn failure into a set of fixable problems, spot what seemed to be missed, and rebuild with those things explicitly in place. People usually call this progress—sometimes maturation, the move from early experimentation to institutional reliability.

Annelise challenges that assumption directly. What looks like progress is often a system becoming more legible to existing institutions, not more structurally different.

The assumption is understandable. Failure meant immaturity, not a broken structure. Something broke because it lacked the components to operate at scale. So the answer seems simple: add oversight, formalize governance, meet regulatory expectations, and plug the system into the frameworks that have long managed financial risk.

To an extent, that view is right—but it leaves out what matters. The response is not neutral. It is shaped by the same forces that shaped the system before it failed. "The response to failure is shaped by the same forces that caused it." — Annelise Those forces don't vanish just because failure happened. They stick around and steer the rebuild.

After events like FTX, the usual takeaway is not that the entire system is rotten, but that it wasn't structured enough. The problem becomes a lack of oversight, a lack of controls, and a lack of the institutional tools that traditional finance built over decades. These include a lack of boards, a lack of reporting standards, and a lack of fiduciary discipline.

Those points dominate the analysis. And from that, the response follows: more structure, more regulation, more integration with the existing financial system. What gets created then is not something new; it's the old thing translated into a different technological environment.

You see this in tokenization, which is often presented as a defining innovation after a collapse. She points to tokenization as a clear example of this dynamic. It is often framed as transformation, but in practice it shows how new infrastructure can reinforce existing structures rather than replace them.

Tokenization involves placing traditional financial assets—funds, securities, real estate—onto blockchain infrastructure. It speeds up settlement, cuts friction in transfers, and lowers the bar for who can take part. Those changes fix real inefficiencies that have long existed in traditional finance, but they don't change the basic structure. The assets stay the same; the ownership rules stay the same; and the rules that govern issuance, transfer, and management still apply.

What changes is the medium used to access and move these assets, not the relationships that make them work. The system gets faster and more accessible and more efficient, but it doesn't become structurally different in ways that would change how it behaves under stress. We often miss that because tokenization's benefits are visible right away: faster settlement, lower costs, and broader participation.

Tokenization feels like transformation. But real transformation requires more than efficiency. "Changing the infrastructure doesn't necessarily change the underlying system." — Annelise It needs a change in who controls things, how risk is managed, and who is held responsible. In many tokenized systems, these things stay aligned with the old structures even as the user interface looks modern. Control keeps its role; institutions still manage funds; custodians still safeguard assets; and regulators still set the lines for the system that runs. The blockchain does not remove these roles; it just puts them in a different operational context.

That repositioning can look like decentralization. But it often just redistributes functions inside a system that remains centralized. That's not a failure of innovation. It's the environment in which innovation gets used. Financial systems don't float free, but they sit inside legal, regulatory, and institutional frameworks that limit what is possible. To scale inside that world, systems must align with those limits. And alignment means taking on many features of the existing system. So traditional financial firms showing up isn't an accident; it's structural.

From Annelise's perspective, this is not a shift away from traditional finance, but a convergence toward it. The system adapts in ways that make it more compatible with existing institutional expectations. Firms that have spent decades inside regulated environments know how to manage capital at scale, how to handle compliance, and how to build trust under scrutiny. Their presence brings stability and expectations around control, accountability, and risk management that shape where the system goes. Those expectations are concrete. They decide how assets are issued, how they are held, how they move, and when they can be accessed.

As blockchain systems merge with these rules, they reflect them. The system becomes more legible and more compatible with existing finance. But compatibility introduces constraints. They can put money in, but they don't set the rules for how that money is used. The system takes on more participants and keeps operating the same way inside. That need pulls systems toward being compatible with the old ways. If you want to work with regulated institutions, get institutional capital, and obey the law, you end up molding your design to match.

As bridging becomes central, new choke points appear. The underlying tech can stay distributed, but the access points and pathways become more centralized. You get layers. One layer looks decentralized, while another layer funnels through centralized hubs. That layered design is not inherently unstable—it is familiar. We've seen this before. When people improve a system, they also add dependencies and control points. Now it's often done on purpose, to make the system steadier and easier to scale. The goal is to avoid past failures. The outcome is a system that works better within its world. But the basic dynamic remains: structure shapes behavior; control shapes outcomes; and dependencies set limits. Those things don't vanish after a failure, but get rearranged. And when they're rearranged, they often become clearer.

A system after a collapse can look very different in some ways and still run by the same logic—unfamiliar words, new tools, new players—but capital, control, and risk still behave the same way. That is not accidental; the constraints have not changed. If those constraints remain, systems will continue to evolve to reflect them.

Does that mean progress is illusory? No. Progress happens. But it happens inside boundaries that define what's possible. Inside them, systems can grow faster, easier to use, and smarter. They're unlikely, though, to change in ways that remove the dynamics that caused earlier failures. That's the tension.

People want to innovate, and they also must conform. Innovation pulls forward while conformity pulls back. The path is not linear—it is cyclical. Systems grow and hit limits. They then get rebuilt to fit their environment, as each round learns from the last. But each round also carries forward the same structures.

So, responses to failure often reinforce continuity. They stabilize things and make the system tougher, but they don't change the underlying conditions. As long as those conditions remain in place, the system keeps moving in the same pattern. This phase is hard to spot because it feels like fixing things, not repeating them. The language shifts and new people join; hence the systems become more structured and easier to read.

From the outside, it looks as if lessons have been learned. The weak spots were addressed, and the system seems improved. And it is improved with new governance, alignment with rules, and institutional practices. Those make the system more stable in obvious ways: risk becomes easier to spot, accountability gets clearer, and ways to handle failure are built in. These are actual changes that matter, especially when people are watching. Stability is not transformation. "What looks like progress is often alignment with existing systems, not transformation." — Annelise

What changed is how well the system works within its environment. The environment itself did not change. Laws, capital flows, institutional habits, market expectations—they're still there. The system adapts, but it does not remove those limits because it can't.

This distinction sits at the center of Annelise's argument. Rebuilding improves function within existing constraints, while redefining would require those constraints to change. "Rebuilding improves function within constraints. It doesn't remove those constraints." — Annelise

That's the difference between rebuilding and redefining. Rebuilding works within the old boundaries. It tightens processes, adds safeguards, and lowers the chances of certain failures. Redefining would require a change to the rules of the game. It would mean different governance, different capital flows, and a different distribution of control.

Most responses choose to rebuild, not because redefining is bad, but because it's constrained. Systems don't sit alone; they live in a wider world that tells them how to behave. To scale, to get capital, to plug into finance, line up with that world. That alignment steers the evolution. Compatibility enables growth but limits the scope of change.

Systems can look new but still behave in old ways. They use better tech, get faster, and open access to more people, but they keep operating under centralized control, institutional oversight, and reliance on the usual trust mechanisms. That doesn't wipe out the value of the gains. It just puts them in context. Progress happens within limits, and innovations get bent by those limits. The result is a system that's more refined but not different in how it would prevent past failures.

The cycle goes on. Not because people cannot improve, but because improvement doesn't happen in a vacuum. It happens inside a framework that sets the range of outcomes. Thus, the response to failure becomes a continuation of what came before. It brings lessons, protection, and more resilience in some areas, but it doesn't transform the deep structural conditions that shape behavior under stress.

People often miss this because new tech and fresh stories make the system look like it has moved to another phase. But beneath the surface, the same relationships persist: capital stays concentrated, control finds central spots, and risk gets managed in ways that favor stability over spreading power. Those are not bugs. They are structural features.

And in her view, those constraints are the part that rarely changes. Until they change, patterns repeat.

But how a system responds doesn't determine what it becomes. What matters is how people behave within that response.

Stress Reveals Truth

If pressure shapes systems and collapse reveals structure, then behavior defines them. Not ideas, design, or narrative—but how people act when given the opportunity. "Behavior does not follow intention. It follows incentives." — Raj

This is the way Raj framed it. Not as theory, but as something that shows up repeatedly in real market conditions. His focus wasn't on what systems do, but on what people do when given the opportunity.

And when incentives are built in deep, behavior is the clearest sign of what a system really is. Stress is revealing because it strips away abstraction and forces a system into its basic functions. It shows what people prioritize when rules tighten and opportunities shrink. When consequences get closer to actions, behavior gets harder to hide. It shows not what the system says it values, but what it rewards.

In crypto, this often gets missed. Busy, speculative periods are often dismissed as noise. Meme coins, fast token launches, quick trading cycles, sudden spikes of attention—these are called distractions from the "real" work of building infrastructure and use cases. People treat them as proof that the industry is immature, still chaotic. That takes away an important point. What looks chaotic on the surface can be organized underneath.

Raj pointed to moments like these as examples of alignment, not distraction. What looks chaotic is often behavior responding directly to what the system makes possible. Memecoins do not emerge by accident. They respond to incentives. When launching a token costs little, when you can distribute it fast, and when attention can be turned into money quickly, people go after that. It's not irrational, but it's aligned with the system.

Under stress, that alignment shows up more. When markets speed up, liquidity moves, recent stories appear, and decisions happen fast. Capital flows to immediate opportunities, not long-term plans. In those moments, design matters less than incentives. People follow what the system allows and rewards.

That makes the behavior diagnostic. Look at token floods on networks like Solana, where thousands are being created daily. Many call that excess, but it's also a genuine test. It pushes infrastructure by straining throughput and latency in ways lab tests cannot. It finds bottlenecks and shows what breaks under real demand.

As Raj observed, these moments act as stress tests. For him, the value isn't in judging the activity, but in observing what it reveals. Stress doesn't distort the system. It exposes it. Not just for the tech, but for everything around it. Real-time stress testing is not theoretical. A planned simulation controls variables and guesses outcomes. Real stress introduces unpredictability—where technical strain meets human behavior and financial pressure meets psychology. You learn whether a system only works on paper, or how it behaves when it's pushed past its limits.

Meme coins are not just a sideshow. They are a mechanism that pulls attention. They speed up participation and compress time, and by doing that, they expose how the system actually works. One big pattern in these settings is that attention drives behavior. People don't put capital only into utility or fundamentals. They follow visibility, momentum, and narratives, too. Assets that attract attention draw liquidity—and that liquidity reinforces attention. The loop runs fast and can crash fast too.

This isn't unique to crypto. You see it in stocks, media, anywhere visibility affects value. But crypto amplifies it through lower barriers to entry, faster interactions, and tighter feedback loops. Things move quickly and reveal patterns that slower systems hide.

This is a point Raj returns to repeatedly. Systems can be built around principles, but behavior aligns with what is rewarded, not what is intended. This shows that incentives override ideology. "The system is not defined by what it claims to be, but by what it consistently produces under pressure." — Raj

A system might be built around decentralization and transparency, but people chase opportunity. When incentives align with those principles, they help. When they don't, behavior drifts. That drift isn't a moral failure. It's a mirror of the system.

People act rationally within given constraints. "People act rationally inside the rules they have." — Raj

They chase returns, cut risk, and follow signals. If speed is rewarded, they act fast. If attention is rewarded, they chase visibility. If asymmetry beats stability, they take risks. Stress makes that truth visible.

The system is not defined by slogans, but by outcomes. And outcomes come from incentives meeting behavior. For infrastructure that matters a lot. High-volume activity—speculative, manipulative, or novel—forces systems to prove themselves. It tests onboarding, throughput, latency, and UX in ways models miss. It shows friction points and failure modes. It points to what needs fixing. That's why frenzies often speed iteration.

Builders watch how systems perform under load, find failures, and tweak things. UX gets better, performance improves, and infrastructure gets tougher. Not just because someone planned it, but because the system faced actual conditions. Behavior during those times reveals deeper things by showing where value sits, how fast money can move, and how people react to uncertainty.

Difficult to see in calm periods, stress reveals it. What follows from that is not just observation, but adaptation. From Raj's perspective, this is where the system stops being abstract and becomes operational. "Behavior doesn't just reflect the system. It reshapes it." — Raj

When patterns in behavior become consistent, systems adjust around them. High-volume activity is not ignored; it is studied. Developers respond to it. Infrastructure develops to handle the kinds of actions users actually take, not the ones designers originally imagined.

If trading dominates, systems optimize speed and execution. If attention drives participation, interfaces shift to capture and direct that attention. If capital moves quickly, settlement and liquidity pathways adapt to keep up. These changes are not planned in isolation. They are reactions to sustained behavior.

Over long stretches of time, this creates reinforcement. Behavior shapes infrastructure, and infrastructure makes that behavior easier, faster, and more central. What began as one pattern among many becomes the dominant one. Not because it was chosen, but because it was repeatedly expressed and supported.

Those adjustments eventually stop being reactive and start becoming structural. When a system repeatedly adapts to the same kinds of behavior, it embeds those patterns into how it operates. What started as a response becomes part of the system's design. Pathways that once supported a range of uses narrow around the most common ones. Interfaces, liquidity flows, and technical optimizations all converge toward the behaviors that appear most often.

Raj describes this as the point where repeated behavior stops being optional and starts becoming structural. This is how systems develop path dependence. Early patterns do not just influence the system temporarily; they shape the direction of its evolution. The more a system supports a behavior, the more efficient it becomes. And the more efficient it becomes, the more it is repeated. That repetition creates lock-in.

Once a behavior is supported by infrastructure, reinforced by liquidity, and validated by repetition, it becomes difficult to displace. Alternatives may exist, but they operate at a disadvantage. They lack the same support, the same speed, and the same visibility. Over time, the system reflects its dominant behaviors, not its full range of possibilities.

This does not happen because anyone decides it should. It emerges from the interaction between incentives, behavior, and adaptation. The system becomes easier to use in certain ways and less effective in others. That imbalance shapes participation. And once that shape is set, the system does not need to be directed. It continues in the same direction because everything inside it has been tuned to support it.

The supposed split between "serious" use and "speculative" activity matters less than people think. They're in the same system, have the same incentives, but differ in how they show up. Speculation is simply more visible—it operates on short horizons and provides faster feedback. But attention, momentum, and incentive alignment exist at every layer, even in the parts called fundamental. What's left is recognition.

The system's behavior is not separate from its structure. It's the clearest expression of it. And once it shows itself, the system is harder to misunderstand. Not simpler. Just revealed. It reflects what users do. That's why trying to separate "serious" from "speculative" often misses the full picture. Both come from the same structure, but the gap between them is time horizon and visibility, not essence.

Long-term builds and short-term speculation run on different scales, but they chase the same thing: value inside the system's limits.

That movement isn't planned. It emerges naturally, is strong, and changes who wins and who loses. It guides adoption across the whole ecosystem. In reality, behavior follows incentives. That difference matters because it explains why well-designed systems still produce surprises. And it explains why calm periods can switch into sudden shifts. It also explains why patterns repeat even as the system changes.

Incentives stick. They show up repeatedly, maybe in new clothes, but doing the same work. This sits at the core of Raj's view. The system is not defined by what it claims to be, but by what it consistently produces under pressure. What this shows is not just how people behave, but how systems settle into patterns.

Incentives do not need to be restated to remain effective. Once they are embedded, they continue to guide behavior across cycles. New technologies, new narratives, and new participants enter the system, but the underlying drivers remain. They reappear in different forms, attached to unique assets, but producing the same kinds of outcomes.

Therefore, behavior is the most reliable lens. It is not filtered through intention or interpretation, and reflects what the system enables in real time. What people repeatedly choose to do is not random, but it is aligned with what the system rewards. That alignment defines the system.

It is not what the system claims to be—it is what it produces.

Because in the end, the system isn't defined by what it was designed to be—it's defined by what people consistently do within it.

The Pattern

Patterns are hard to spot not because they are complex, but because they are familiar. They do not appear as unknown facts or demand attention. They emerge through repetition. The same dynamics show up again and again. On the surface, things look different, but underneath, the structure lines up.

Each case has its own story, its own people, and its own context. So, we call it unique. We focus on the differences, parse the specifics, and build explanations that fit that moment. What gets missed is not data, but the persistence of structure.

They set limits on what a system can do and shape its path in ways that are not immediately visible. Inside those limits, systems adjust—shifting emphasis, making tradeoffs, and responding to pressure. But they usually follow paths set by the same constraints.

That's where patterns appear. Seeing that means shifting your view away from single events and toward the conditions that make those events possible. Events are visible, but they are not the system. They're outcomes, not causes. You can analyze them closely, but that analysis stays tied to the moment. To understand more, ask how those conditions were made. How did the system move to a place where those outcomes were likely? And where else might the same conditions be forming?

That's where the sequence this book looks at matters. Not a neat story, but a process that repeats. Systems feel pressure as they scale—more users, higher expectations, greater demand. Pressure often signals that the system is being used beyond its original design. People and capital notice it, and it is being used beyond its original design. But pressure brings limits, and those limits must be managed.

Systems adapt to manage these limits. Adaptation is not neutral—it forces tradeoffs, choices about speed versus decentralization, about cost versus security, and about access versus control. Each choice fixes one problem and creates new dependencies.

Over time, those choices stack up. The system becomes more complex—not just technically, but in the relationships that sustain it. When complexity builds, the structure grows harder to see. If a system runs fine, you don't notice the underlying rules. People use interfaces and apps, functions work, transactions go through, and liquidity looks available.

It appears stable. Those are real—but they are surface signals. They show behavior, not the support that makes behavior possible. Eventually, those assumptions can stop matching reality. This misalignment doesn't always cause immediate failure because the system compensates, adapts, and keeps acting the same even as strain builds.

That looks like resilience, but it also means the system leans more on assumptions that are fraying. What looks strong on the outside can mean the system can only run under narrower and narrower conditions. Collapse feels sudden, but that reflects perception—not the underlying process.

Breakdown is a process. The system reaches a point where it can't maintain the split between how it behaves and how it's built. At that moment, the assumptions it relied upon fall away. The system must operate on verification instead of belief, which is when the structure comes into view. Seeing the structure doesn't cause failure; it reveals it.

The relationships, dependencies, and limits were there all along. They just weren't tested fully. When the conditions change, the system does not become something else—it becomes clearer, and after clarity comes rebuilding. Systems are not abandoned when they break. They get reorganized, and people look for what went wrong and try to stop it from happening again.

Governance becomes more formal, and oversight ramps up. The system becomes more aligned with existing frameworks. Those steps can make it work better under scrutiny and cut certain risks, but they happen inside the same constraints that existed before.

Because constraints persist, key structural elements remain. Control stays concentrated where coordination or compliance needs it. Capital keeps moving through channels that fit institutional rules. And incentives still push behavior in familiar directions. The rebuilt system may look different in design and operation, but it often keeps the features that show up under pressure.

This is why patterns repeat across cycles. Each round brings changes, but the deep dynamics stay the same. Pressure mounts as things scale, while adaptation adds complexity and dependence. Perception and structure drift apart, though collapse fixes that drift. Reconstruction puts the system back into a form that matches its environment.

Through it all, behavior reflects incentives that persist at every stage, giving a steady signal of how the system actually works. That can be hard to accept because it clashes with the idea that learning equals transformation. Learning occurs—systems identify weak points, people adjust behavior, and fixes are implemented. But learning runs up against what the system can handle. It improves processes, tightens controls, and makes things more visible. It doesn't always change the structural conditions that caused the original mismatch.

Those conditions are accounted for in how the system interacts with the world. Capital wants returns. Institutions want stability. Participants want opportunities. Regulators want control. These forces rarely align and instead pull the system in different directions. The system changes as a result, finding temporary balances that keep it running but never fully settling the competing demands.

In that light, the idea of a final state looks unlikely. A "mature" system suggests stability and alignment between structure and behavior. But in systems driven by innovation and capital flows, stability is fragile, predictability is limited, and alignment doesn't stick. The system can get close for a while, but it can't stay there forever because the surrounding forces keep shifting.

You notice the same forces at work across different cases, shaping outcomes in ways that are predictable in structure even if the details differ. That doesn't remove uncertainty, but it gives you a clearer way to deal with it. Clarity here isn't about making the system simpler. It's about spotting the relationships that define it. Like pressure causing adaptation, which then creates dependency and opens the door to divergence, and how divergence gets patched by collapse and rebuilding.

It's also noticing how behavior mirrors those dynamics, giving a running signal of how the system works. The signal is messy, and it moves across time scales and activity layers. Long-term trends sit next to short-term swings while structural shifts happen alongside fleeting behavior. Telling these apart means paying attention to scale and context. What looks big in one frame can be small in another.

Still, some patterns show up no matter what the scale. Capital concentrates; intermediaries appear; and unexamined assumptions crop up again and again. These things aren't always bad, but they set up gaps between how the system looks and how it's structured.

That gap is the critical point—where outward behavior no longer matches underlying reality. If the gap stays manageable, the system keeps working, but if it grows past a limit, the system has to reconcile the difference. And that reconciliation often looks like collapse. This does not prevent collapse, but it helps identify when boundaries are near, and moves the focus from reacting to events toward watching conditions, from studying outcomes to understanding the processes that produce them.

The change is subtle, but it matters. It shifts how people engage, how they read signals, and how they judge risk. It builds a kind of awareness that doesn't require precise predictions. Instead, it lets you spot patterns that make certain events more likely. That doesn't erase uncertainty, but it changes how you experience it.

It also changes how you think about progress. Progress is no longer a march toward stability, but an evolution within constraints. There can be gains in efficiency, access, and function, and simultaneously the old dynamics keep shaping how those gains appear. This doesn't make innovation any less real; it just puts it in context.

Innovation occurs within a system that both adapts and is constrained, so its effects must be read against the forces that steer the system. In the end, this isn't a deterministic view, but a structured one. It accepts variability and places it within a recognizable pattern. The details of each cycle will differ, although the underlying dynamics stay the same and can be observed and interpreted.

This understanding isn't a finish line. It won't resolve the system's tensions or stop the forces that drive change. But it gives a way of seeing that leans less on tidy stories and more on structure. It helps you understand the system over time, not just in isolated moments. And by watching behavior across cycles and conditions, the pattern becomes clearer. Not something forced from outside, but something that shows up from inside the system.

What separates a system that's understood from one just being watched is not how much data you have. It's the ability to spot continuity in variation. Systems under pressure will always change. They create novel forms, new mechanisms, and fresh stories that match their current state. Those things are necessary. They let the system adapt, stay relevant, and add capabilities. But they also make the surface move, and that movement hides what stays the same underneath.

Structure ties things together. It lets you place events inside a larger process, joining what look like separate developments into a running sequence. That doesn't make single events unimportant. It just changes how you read them. They become signs of underlying conditions, points in an ongoing process.

Moving from event-driven interpretation to structural understanding changes how you perceive the system. Disruption often turns out to be the clearing of built-up conditions. So-called innovation is often a reshuffling of existing parts. And uncertainty comes from forces that haven't been reconciled.

The role of the participants becomes clearer in this picture. Participants aren't outside the system; they're part of it. Their choices and interpretations shape its path. They put up capital, spin stories, interact with infrastructure in ways that affect behavior. They're also shaped by the system's structure, by the incentives and constraints it sets.

That dual role, shaping and being shaped, creates a feedback loop at the heart of the pattern. Behavior changes structure, structure changes behavior, and their interaction moves the system. The loop gets pushed by outside things like rules, tech shifts, and macro changes, but its core dynamics stay steady, giving continuity across cycles. It's in that continuity that the pattern shows most clearly, not as a fixed script, but as relationships that produce similar outcomes when conditions repeat.

Pressure makes systems add complexity and dependencies. This builds up divergence between perception and structure. If the divergence crosses a threshold, collapse follows. Then reconstruction happens, fitting the system better to its environment. Throughout, behavior reflects the incentives that persist at each stage. This sketch won't cover every detail, but it captures the core dynamic. It shows how different parts interact and how those interactions lead to outcomes that vary but are consistent.

The upshot isn't that everything is predetermined; it's that outcomes are conditioned. The system doesn't lead to a single end, but moves within a range set by its structure. Within that range, some outcomes become more likely under certain conditions. Spotting those conditions won't guarantee any unique result, but it explains why certain results become more probable.

That difference matters. Prediction tries to pin down specific events, while understanding looks at the conditions. In complex, fast-changing systems, prediction has limits. There are too many variables that interact quickly. Outside forces pull and push, rendering precise forecasts nearly impossible to predict.

Understanding works on another level. It doesn't say exactly what will happen, but it tries to spot the conditions that make some outcomes more likely and others less so. That kind of understanding adapts and lets you change course when conditions shift. It gives a way to make sense of new information, but it doesn't remove uncertainty. Though it makes uncertainty feel less like randomness. Instead, it shows variability within a system that follows consistent dynamics.

Those dynamics change too and move as the system moves. New technology opens new possibilities and changes how pressure builds and gets managed. Rule changes reshape the environment and how systems are put together and how they fit with older rules. When capital moves around, resource flows change. That affects how fast things can scale and how they respond to stress.

These things shape behavior. But they do so inside a structure tied to the same basic relationships. That mix of change and continuity is the system's character: not totally stable, not totally random. It sits in the middle and forces both limit change and makes it possible. So, the surface can shift a lot while the deeper structure stays familiar.

Getting that duality helps make sense of how the system develops. It explains why new versions can look totally different but still give familiar results. It shows why fixing one problem can create a weakness somewhere else, and it shows why cycles keep coming back even as the system gets more advanced.

This way of explaining things doesn't just blame people or single choices. People matter, sure, but their actions happen inside a system that narrows what they can do. If you only look at individuals, you miss the structural conditions that push certain actions and outcomes. That doesn't let people off the hook. But it puts actions in a larger context of incentives, constraints, and the information people had. Seeing that gives a fuller picture, one that includes both personal choices and the system's pull.

What this ultimately shows is not a fixed outcome, but a consistent process. Systems do not move randomly. They evolve within constraints, shaped by pressures they cannot avoid and incentives that do not disappear. While each cycle introduces variation, the structure guiding that variation remains recognizable, which is why the same dynamics reappear even as the surface continues to change.

Understanding this does not grant control—it changes perception. It shifts attention away from isolated events and toward the conditions that make those events possible, replacing the need for precise prediction with the ability to recognize when certain outcomes are becoming more likely, and making it harder to mistake novelty for transformation.

The system does not need to be explained again; it needs to be seen.

About the Author

Jamil Hasan is the founder of Crypto Hipster Publications and host of the Crypto Hipster Podcast, a platform built around a single idea: where builders talk freedom, not price.

Across 580 conversations, he has focused on founders, entrepreneurs, and independent creators—the people actively building the digital economy from the ground up. His work does not center on market cycles or short-term narratives, but on the motivations, risks, and convictions behind those shaping new systems.

His approach is rooted in separating the signal from the noise. In an industry often driven by speculation and surface-level commentary, he prioritizes long-term thinking, first-principles discussion, and authentic perspectives.

Before fully committing to the digital asset space, Jamil built his career across financial services and data-driven initiatives, including work connected to blockchain, artificial intelligence, and client-facing strategy. That foundation informs his ability to translate complex technological change into clear, human insights.

After stepping away during a period of intensive medical treatment, he returned with a sharper lens and a more defined mission: elevate real builders, reject noise disguised as insight, and focus only on what endures.

The Pattern represents the next evolution of that mission.

Crypto Hipster Podcasts

A list of podcasts that form the basis of this book from Seasons 6-8 is presented below. All of them can be found at the Crypto Hipster Podcast station wherever enjoy your favorite podcasts.

- Why Decentralization May Face an Existential Crisis During the Trump 2.0 Era and What We Can Do About It, with Josh Bowen @ Astria (3/26/2025)

- Crypto Hipster Presents... Shooting from the Hip! Episode 11: How the Crypto Industry is Successfully Recovering from the FTX Collapse, with Ishan Bhaidani @ SCRIB3 (5/23/2024)

- Crypto Hipster Presents: Reporter on the Ground, Episode 1; Why Today's Tech Entrepreneurs Will Become Tomorrow's Successful Business Leaders, with Annelise Osborne @ Kadena (6/4/2024)

- How to Power the Borderless Global Economy and Bring Web Three to the World, with Raj Parekh @ Portal (6/30/2024)

www.ingramcontent.com/pod-product-compliance
Lightning Source LLC
Chambersburg PA
CBHW021328060726
47591CB00006B/1924